YOUR KNOWLEDGE HAS VALUE

- We will publish your bachelor's and
  master's thesis, essays and papers

- Your own eBook and book -
  sold worldwide in all relevant shops

- Earn money with each sale

Upload your text at www.GRIN.com
and publish for free

Md. Siddiqur Rahman, Salena Akter

Spatial Distribution of Gmelina Arborea in Bangladesh

A Comparison between Climate Requirements and Dendro-Ecological Records

GRIN Publishing

Bibliographic information published by the German National Library:

The German National Library lists this publication in the National Bibliography; detailed bibliographic data are available on the Internet at http://dnb.dnb.de .

Imprint:

Copyright © 2015 GRIN Verlag GmbH
Print and binding: Books on Demand GmbH, Norderstedt Germany
ISBN: 978-3-656-96155-0

This book at GRIN:

http://www.grin.com/en/e-book/298214/spatial-distribution-of-gmelina-arborea-in-bangladesh

Spatial distribution of *Gmelina arborea* in Bangladesh: a comparison between climate requirements and dendro-ecological records

Md. Siddiqur Rahman [1] **and Salena Akter** [2]

[1] *M.S. in Forestry, Institute of Forestry and Environmental Sciences, University of Chittagong, Chittagong 4331, Bangladesh*

[2] *Department of Environment, Ministry of Environment and Forest, Dhaka - 1207, Bangladesh*

Abstract

Depending on which factors control spatial distribution of forest tree species, climate and edaphic are the two most scheming one those act together to locate tree species diversity. Species spatial arrangements determined by these two factors plotted in a Geographic Information System (GIS) showed relevance with what dendro-ecological studies marked for major forest tree species. As forest tree species has very close correlation with climate factors in their distribution and growth, methodological approach regarding using these two factors applied on a native tree species of Bangladesh: *Gmelina arborea* Roxb. Studied species is a promising one growing naturally in evergreen and deciduous forests of Bangladesh. Not only found in natural forests, it has been planted widely in all plantation sites throughout the country. Present study is to identify potential productive area of *Gmelina arborea* in Bangladesh considering climate and soil factors favourable to its growth. Geographic Information System (GIS) is a key tool to plot favourable plantation areas matching temperature, rainfall, soil types and topographic factors for the species studied. Results depicted that *Gmelina arborea* might grow well in Eastern part of the country especially in North-Eastern and South Eastern part of Bangladesh. This may be due to higher temperature and rainfall on that part and favourable sandy loam soil for the species there. These proposed potential plantation areas fall in tropical evergreen and semi-evergreen forest of the country, known to be its natural habitat. By this study it would be easier to identify *Gmelina*'s distribution in forest ecosystems that may determine which silvicultural treatments are suitable for it to grow towards a normal growing stock.

Key words *Gmelina arborea*; distribution; Bangladesh; Climate factors; Edaphic factors

Introduction

Climate is the key factor in determining forest productivity and distribution (Rahman et al., 2015). It is evident that there is a close relation between the climate and the vegetation it supports (Champion et al., 1965). Moreover, climate factors like temperature, precipitation, growing season length, light factors and humidity proved to influence forest potential productivity greatly when soil and topographic factors remain constant (Rahman et al., 2015). Potential productivity of vegetation differ significantly in various Agro-ecological areas, particularly it is lower in the North Western and South Western part of Bangladesh, on the contrary, higher in the Eastern part (Al-Amin and Rahman, 2011b). Vegetation distribution and productive area can be determined by Geographic Information System (GIS). Spatial distribution and shift due to change in climate factors may be visualized using GIS techniques (Rahman, 2012). Thus, GIS is considered as a potential tool for forest management (ESRI, 2001). Once data are entered in a GIS, maps can be displayed showing general species distributions and the area of stands can be calculated (Green and Congalton, 1990). Species distribution in Bangladesh depends largely on climate, site productivity and silviculture of itself (Al-Amin and Rahman, 2011a).

Gmelina arborea is a widely grown forest tree species found in rainforest as well as dry deciduous forest and tolerates a wide range of conditions from sea level to 1200 m elevation and annual rainfall from 750 to 5000 mm. It grows best in climates with mean annual temperature of 21–28°C (Jensen, 1995). It is a medium to large tree that reaches 35 m in height and more than 3 m in diameter in natural stands in tropical and subtropical regions of Asia (Dvorak, 2003). It is a fast growing species member of *Verbenaceae* family (Chudnoff, 1979) that has become a major international timber species over a wide range of sites in the tropics. The species occurs naturally from latitudes 5° to 30° N and longitudes 70° to 110° E. Its altitudinal range is approximately 50 to 1300 m in areas with distinct dry seasons in the countries of Bangladesh, Cambodia, China (Yunnan and Kwangsi Chuang provinces), India, Laos, Myanmar, Nepal, Pakistan, Sri Lanka, Thailand, and Vietnam (Lamb, 1968; Dvorak 2003, Lauridsen and Kjaer, 2002; Duke, 1983; Hossain, 1999). It grows almost all over Bangladesh (Hossain, 1999). *Gmelina arborea* grows well on deep, loamy, clay loams, calcareous, and moist soils with optimum rainfall from 1800 to 2300 mm per annum (Tewari, 1995; Lauridsen and Kjaer, 2002; Wijoyo, 2000; Espinoza, 2003). It grows best on deep, well drained, base-rich soils with pH between 5.0 and 8.0. Growth is poor on thin, highly leached acid soils (F/FRED 1994). And now it's widely planted in many countries of the lowland tropical zone because of its ability to survive, coppice and yield seed early in various climatic conditions (Sulaiman and Lim, 1989). In the natural forest, the species is usually found scattered and in association with other species. It is found in evergreen forests in Myanmar and Bangladesh and in relatively dry mixed deciduous forest types in Central India.

Plants originating from plantations often perform very well compared to those originating from natural forests. This can be due to (i) a positive selection during thinning in the plantations, (ii) a result of lower inbreeding in the plantations, or (iii) a positive response to adaptation to local conditions (because the plantings often represent local 'landraces') (Lauridsen and Kjaer, 2002). Therefore, seeking potential productive areas for *Gmelina arborea* is very much important because it is widely planted throughout the country especially in tropical evergreen and semi-evergreen forests extended over Chittagong, Cox's Bazar, Chittagong Hill Tracts and Sylhet. One most valuable plantation species here is Gamar (*Gmelina arborea*). Thus, there is a need to identify the productive zone of this species with respect to climate requirement as future forest might be altered by changing climate.

Materials and Methods

Materials

Present study requires some base information as Bioclimatic factors used in the study were Iso-thermal lines throughout Bangladesh (Rashid, 1991), Iso-Precipitation lines throughout Bangladesh (Rashid, 1991), Seven Soil Tracts of Bangladesh (Banglapedia, 2006) and silvicultural requirements of *Gmelina arborea* form Troup (1921 and 1986), Das and Alam (2001), Luna (1996). Forest map of Bangladesh was adopted from FAO (2007).

Study site

In South Asia Bangladesh lies between 20°34' to 26°38' north latitude and 88°01' to 92°41' east longitude. In the context of physiography, the country may be classified into three distinct regions (a) floodplains, (b) terraces, and (c) hills each having distinguishing characteristics of its own (Banglapedia, 2006). She has a humid, warm, tropical climate. Its climate is influenced primarily by monsoon and partly by pre-monsoon and post-monsoon circulations (Agarwala et al., 2003).

Base maps

The studies were conducted using four base maps viz. the iso lines of temperature, precipitation throughout Bangladesh, seven soil tracts of the country and forest area map of Bangladesh (Figure 1).

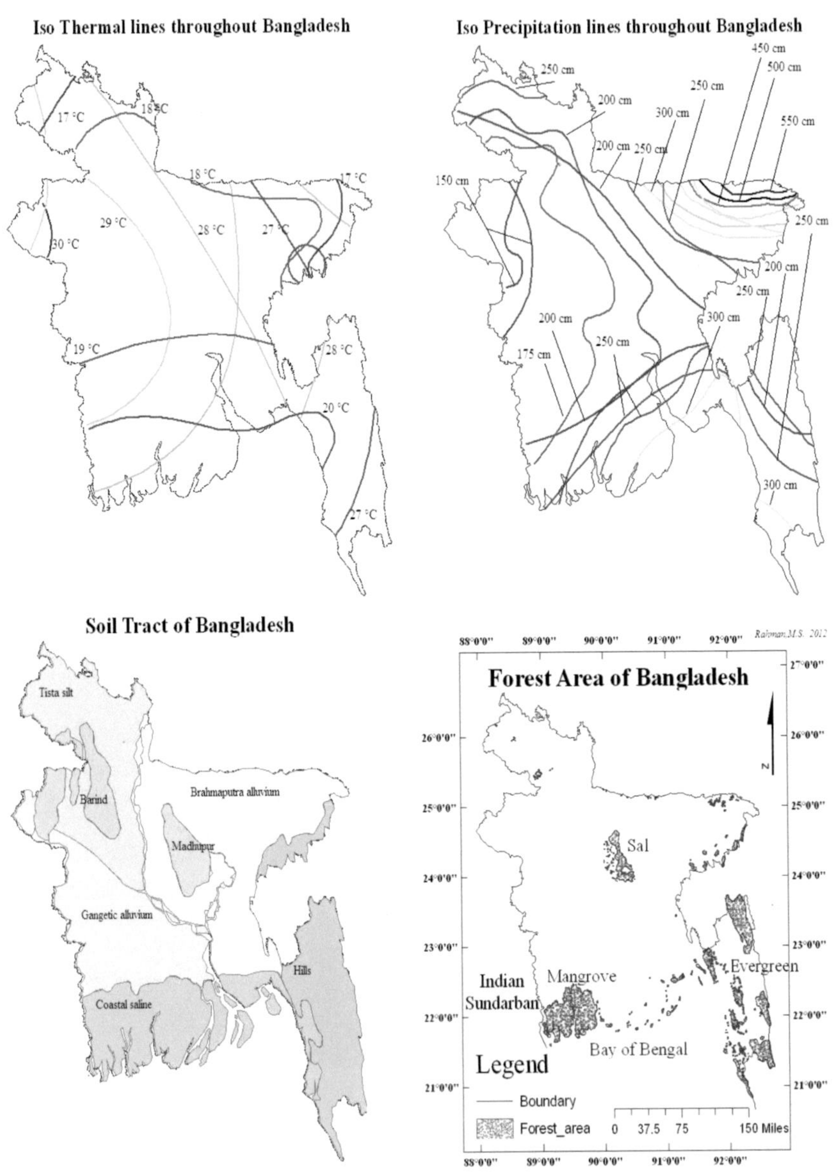

Figure 1. Three base maps used in the study and forest distribution in Bangladesh (Rashid, 1991; FAO, 2007).

Methods

Using Mean Annual Temperature (MAT) and Mean Annual Rainfall (MAR) map, all the different iso-lines of temperatures and precipitations were created. All the MAT and MAR iso lines were buffered by fifteen kilometers on both sides. The buffer tool was found in Proximity sub tool of Analysis tool. Now for each individual tree species all the required temperature and rainfall iso-lines, soil types, tracts were united. This procedure was done by using the union tool in Overlay sub tool of Analysis tool. After uniting each type of polygon say, MAT, MAR, Soil type or Soil tract for the species these polygons were dissolved. Dissolve tool was found from Generalization sub tool of Data management tool. After dissolving all the necessary climatic and edaphic layers for any individual tree species, these polygons were intersected to find the common area between them. Intersection was found from the Overlay sub tool in the Analysis tool. For studied forest species the intersected areas conveyed the spatial distribution of that species in the country. Methodology followed from Al-Amin and Rahman (2011a).

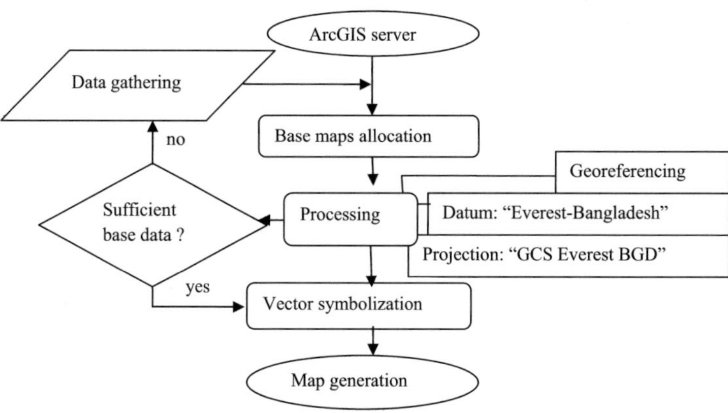

Figure 2. Flow chart of methodology in GIS environment (Rahman et al. 2015).

Results and Discussions

Spatial distribution of *Gmelina arborea* is derived from analyzing the base map information and the resultant map (Figure 3) shows that the species tends to relocate in the eastern part of the country. Both northern and southern part is suitable for the plantation of *Gmelina arborea*. Hence the species may be planted at these locations as forests species try to find their best suitable climatic and edaphic factors to survive.

Spatial distribution of any species tends to locate to its natural productive area. Whether this natural productive area depends on species suited climatic and edaphic factors as favoured. Tree species tends to choose their habitat that most match its criteria. Plantation of forest tree species depends on this criterion. When the site fulfills this, a species may be planted there. This may be useful for economically important forest tree species as choosing best site for plantation might be the first issue as economically and sustainably both. Analyzing species spatial distribution in Bangladesh this may be continued that the species *Gmelina arborea* is very much prone to be planted in Eastern part of the country. Eastern part of the country comprises Evergreen and Semi-evergreen forest types in Bangladesh.

Similar results found when plotting forest and agroecosystem productivity for Bangladesh using only climate factors to be the determining one (Rahman et al., 2015). Forest productivity deemed to be greater in north and south eastern part of the country than western part. The fact showed there was that there may be favourable optimum temperature and high precipitation in eastern region than western region of Bangladesh.

From Dvorak's (2003) *Gmelina* map (Figure 4), it is evident that this species is likely to be better suited in north eastern and slight in south eastern part of the country (Bangladesh). It resembles with present study that finds *Gmelina*'s climate determined distribution throughout the country.

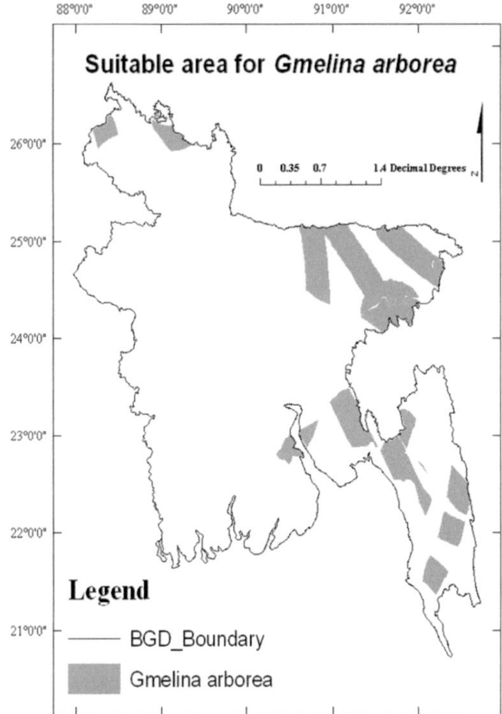

Figure 3. Spatial distribution of *Gmelina arborea* in Bangladesh analyzing their suitable climatic and edaphic factors (GIS map prepared by Md. Siddiqur Rahman).

Thus, comparing between natural silvicultural distribution and anticipated climate-soil suited distribution for *Gmelina arborea* give spatial location at north eastern and south eastern tropical evergreen and semi evergreen forest of the country.

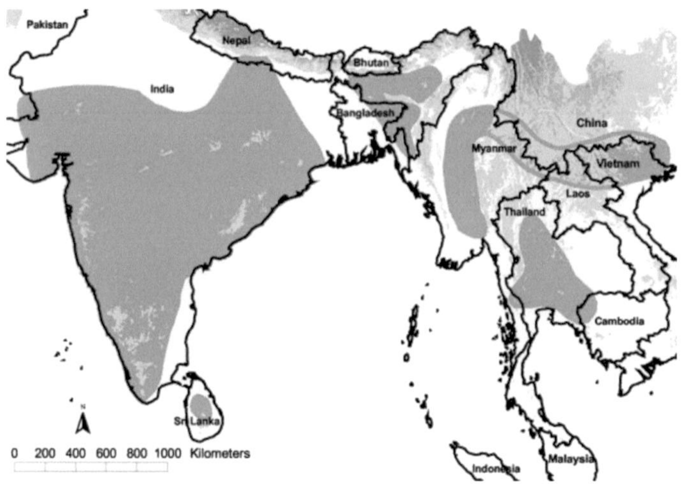

Figure 4. Natural distribution of *Gmelina arborea* in South-East Asia (Dvorak, 2003).

Conclusion

Forest tree species tend to distribute themselves in places that best comprise their limiting factors. *Gmelina arborea* distribution in Bangladesh is very much prospectus. The study found that the species can be planted in eastern evergreen and semi evergreen forest land and other wooded land in Bangladesh. The present study may help plantation managers in the country to find suitable locations to plant with this species.

Acknowledgement

Authors would like to thank IFESCU-USDA funded project at the Institute of Forestry and Environmental Sciences, University of Chittagong premises for supporting with GIS laboratory during the project.

References

Agrawala, S., Ota, T., Ahmed, A. U., Smoth, J., Aalst, M.V. 2003. Development and Climate Change Bangladesh: Focus on Coastal Flooding and the Sundarbans. Paris: Organization for Economic Co-operation and Development (OECD).

Al-Amin, M. and Rahman, M.S. 2011a. Sketching future spatial-temporal distribution of selected forest tree species considering climate change using GIS. Proceedings of the National seminar on "Space technology application for monitoring earth resource, disasters and climate change impacts for ensuring human security and sustainable development", Bangladesh Space Research and Remote Sensing Organization (SPARRSO), 28-29 June, 2011, Dhaka, Bangladesh.

Al-Amin, M. and Rahman, M.S. 2011b. Sketching future forest of Bangladesh considering climate change scenarios, silviculture and productivity of species using GIS. Montpellier, France: IUFRO International Conference on "Research Priorities in Tropical Silviculture: Towards New paradigms?" 15-18 November 2011.

Banglapedia, 2006. National Encyclopedia of Bangladesh. Dhaka, Bangladesh: Asiatic Society of Bangladesh.

Champion, H.G., Seth, S.K. and Khattak, G.M. 1965. Manual of Silviculture of Pakistan. Pakistan: Govt. Press. 540 pp.

Chundnoff, M. 1979. Tropical Timbers of the World. Madison, Wisconsin, USA: US Forest Product Lab., Forest Service, United States Department of Agriculture.

Das, D.K. and Alam, M.K. 2001. Trees of Bangladesh. Chittagong, Bangladesh: Bangladesh Forest Research Institute.

Duke, J.A. 1983. *Gmelina arborea* Robx. Handbook of Energy Crops (unpublished). Available at www.hort.produce.edu/newcrop/duke_energy/Gmelina_arborea.html accesses on 10 March 2012.

Dvorak, W.S. 2003. World View of *Gmelina arborea:* Opporunities and Challenges. In: Recent advances with *Gmelina arborea* (eds. Dvorak, W.S. Hodge, G.R. Woodbridge, W.C. & Romero J.L.). CD-ROM. Raleigh, NC. USA: CAMCORE, North Carolina State University.

ESRI, 2001. ESRI: Creating GIS for a Better World. USA: Environmental Systems Research Institute. www.esri.com/forestry. assessed on 29.01.2011.

Espinoza, J.A. 2003. Site selection, Site Preparation, and Weed Control for Gmelina arborea in Western Venezuela. In: Recent advances with *Gmelina arborea* (eds. Dvorak, W.S. Hodge, G.R. Woodbridge, W.C. & Romero J.L.). CD-ROM. Raleigh, NC. USA: CAMCORE, North Carolina State University.

FAO. 2007. Brief on National Forest Inventory NFI Bangladesh. Forest Resources Development Service, Food and Agricultural Organization of the United Nations, Rome. Available at http://www.fao.org/docrep/016/ap180e/ap180e.pdf. accessed on 12.01.2013.

F/FRED. 1994. Growing multipurpose trees on small farms, module 9. USA: Species fact sheets (2nd ed.). Forestry/ Fuelwood Research and Development Project, USDA.

Green, K. and Congalton, R. 1990. Mapping potential old growth forests and other resources on national forest and park lands in Oregon and Washington. In: Proceedings, GIS/LIS '90, Vol. 2, Anaheim California, USA: November 7-10, 1990. ACSM, ASPRS, AAG, URISA, AM/FM International, 712-723pp.

Hossain, M.K. 1999. *Gmelina arborea:* A popular plantation species in the tropics. FACT Sheet: quick guide to multipurpose trees from around the world. Arkansas, USA: FACT 99-05. Forest, Farm, and Community Tree Network. Winrock International. Available at http://www.winrock.org/fnrm/factnet/factpub/FACTSH/Gmelina%20arborea1.pdf

Jensen, M. 1995. Trees commonly cultivated in Southeast Asia. Illustrated field guide. RAP publications.

Lamb, A.F.A. 1968. *Gmelina arborea*-Fast growing timber trees of the lowland tropics. Oxford, UK: Commonwealth Forest Institute. 31pp.

Lauridsen, E.B. and Kjaer, E.D. 2002. Provenance research in *Gmelina arborea* Linn., Roxb. A summary of results from three decades of research and a discussion of how to use them. International Forestry Review **4**(1): 1-15.

Luna, R.K. 1996. Plantation Trees. Dehra Dun, India: International Book Distributors.

Rahman, M.S. 2012. Climate Change and Forest in Bangladesh: Growth, Survivability, Stress Adoption and Spatial Shift to Forest Species Due to Climate Change. Saarsbrucken, Germany: LAP Lambert Academic Publishers.

Rahman, M.S., Akter, S., Al-Amin, M. 2015. Forest and Agroecosystem Productivity with Respect to Climate in Bangladesh: A Climate Vegetation Productivity Approach. Forest

Science and Technology. http://dx.doi.org/10.1080/21580103.2014.957358

Rashid, H.Er. 1991. Geography of Bangladesh. Dhaka, Bangladesh: University Press Limited.

Sulaiman, A. and Lim, S.G. 1989. Some Timber Characteristics of *Gmelina arborea* Grown in a Plantation in Peninsular Malaysia. Journal of Tropical Forest Science **2**(2): 135 – 141.

Tewari, D.N. 1995. A monograph on Gamari *(Gmelina arborea Roxb.)*. Dehra Dun, India: International Book Distributions. 125 pp.

Troup, R.S. 1921. The Silviculture of Indian Trees. Oxford, UK: Clarendon Press.

Troup, R.S. 1986. The Silviculture of Indian Trees. Oxford, UK: Clarendon Press.

Wijoyo, F.S. 2000. A Study of Genetic Parameters of *Gmelina arborea* Roxb. In: Thailand Growth in 5 Countries and Financial Consideration for Operational Deployment of *Gmelina arborea* in Indonesia. Thesis of Degree of Master of Science. Raleigh, USA: Forestry Department. NCSU.